Wild Plants of Malta

Hans Christian Weber

Wild Plants of Malta

Publishers Enterprises Group (PEG) Ltd

Published by
Publishers Enterprises Group (PEG) Ltd,
P.E.G. Building
UB7, Industrial Estate,
San Gwann SGN 09, Malta

E-mail: contact@peg.com.mt
http://www.peg.com.mt

© Hans Christian Weber, 2004

All rights reserved. No part of this publication may be reproduced,
stored in a retrieval system or transmitted in any form or by any
means, electronic, mechanical, photocopying, recording, or otherwise,
without prior permission in writing of Publishers Enterprises Group (PEG) Ltd.

First published 2004

ISBN: 99909-0-370-0 (paper-back)
ISBN: 99909-0-371-9 (hard bound)

Printed by P.E.G. Ltd, Malta

Acknowledgments

There are some individuals who deserve a special mention. I would like to thank Dipl.Biol. Bernd Kendzior for herbarium searches and picture 108, as well as Dr. Stephan Imhof for copyediting the English text. I would also like to thank Joseph Buhagiar for the stimulating comments made at the very early stages of this project, and Anthony R. Callus for his helpful remarks and corrections. It is a pleasure for me to have Publishers Enterprises Group (PEG) Ltd as publishers of my book.

No contribution to a book of this type is more appreciated than the interesting evening talks and discussions with John and his brother, Hergen, Silvana and the other people at the Palace in Malta!

Last, but by no means the least, I express my sincere thanks and love to my wife Ulla for her constant appreciation.

Contents

Preface 9

Introduction 11

Wild Plants of Malta 15

References 135

Index of Scientific Names 137

Index of English Names 141

Index of Maltese Names 143

Preface

This book is meant to serve as a presentation of Malta's natural beauty. Perhaps it is not accidental that I, as a foreign botanist, felt obliged to research and compile the subject matter of this book. What appears to be "usual" to you, the citizens of Malta, is "special" to me. The main aim of this book, therefore, is to make everybody aware that you live in a botanical paradise – you are the owners of a Mediterranean pearl. By showing Malta's plant diversity, I hope to further intensify your relationship with plants and stimulate responsibility and appreciation for wildlife and nature. During my field trips I had several occasions to make friendly face-to-face contact with Maltese people, and soon learned that not only Malta's plants are lovable. Therefore,

I would like to dedicate this book to you, the people of the Maltese Islands.

I also wrote this book for tourists visiting Malta, encouraging them to visit places like Chadwick Lakes, Mistra Valley, Wied Babu, the Victoria Lines, Għajn Tuffieħa, Selmun Bay, the Red Tower, the White Tower and other locations in Malta. I would also suggest visits to Gozo and Comino: a cliff hike between Xlendi and Ta' Ċenċ is a highlight of country walks through scenic areas. Through these country walks you will be amazed by the diversity of flowering plants. One should always be aware that there are private sites, and since Malta is small, it is even more important to follow the rule: leave only footprints on paths, and only take pictures and impressions.

Hans Christian Weber

Introduction

In all, the three main islands of Malta, Gozo and Comino have a surface area of approximately 316 km^2. It is most remarkable that on these small islands there is about half the number of flowering plants species as in the whole of Germany! One reason for this biodiversity is Malta's geographic position: the Maltese Islands are situated at the centre of the Mediterranean Sea, 95 km south of Sicily and 290 km from the north African coast. Long ago there existed a land connection between Malta and Sicily. In fact, the floras of both islands are closely related. Certainly, the particularly old history of Malta with its multicultural influence over thousands of years has also played an important role in the formation of the island's present day natural and environmental heritage.

The main reason for Malta's remarkable biodiversity is its interesting and obvious geology, with its formations of Upper Coralline Limestone, Greensand, Blue Clay, and Globigerina and Lower Coralline Limestones. Habitats like beaches, rocky coasts or cliffs (see fig. 001) are common occurrences, as are the different characteristic vegetation types of the steppes, maquis and garigues (fig. 002). In contrast, *widien* can be described as dry or temporarily wet open or sheltered valleys (fig. 003) or locations which naturally hold watercourses or even lakes for a longer period of time during the year. Some plant species are strictly adapted to one of the special habitats and, therefore, are often endangered. The flora of Malta includes a number of endemic species: these plants only occur in Malta, and are not to be found anywhere else in the world. Other species of plants, especially those introduced to the islands from other countries, are known to grow just about everywhere. In this book, both rare and common species of wild plants (Gymnosperms and Angiosperms) from all natural habitat types are considered. Agricultural species and those planted for ornamental purposes are not included.

Since a picture is indeed worth a thousand words, and in order to save space for photographs, I have kept descriptions as short as possible and mentioned the scientific, the English and the Maltese name of the plant, whilst providing some information regarding habitat, distribution and the most prominent flowering time. On certain occasions I have added short comments in order to facilitate picture interpretation and include other interesting facts regarding the plant species. The order in which the species are presented follows the actual plant systematics.

Although many other interesting plants of the flora of Malta could not be considered and included here, I am convinced that this book will give a comprehensive and representative cross section of Maltese wild flowers.

Wild Plants
of
Malta

PINACEAE

Family:
Pinaceae

004, 005

Pinus halepensis

English:
Aleppo Pine

Maltese:
Siġra taż-Żnuber, Prinjoli

Habitat/Distribution:
Different sites, common

Comments:
Re-introduced and cultivated. Cones all year round

004

005

17

CUPRESSACEAE

Family:
Cupressaceae

006

Cupressus sempervirens var. horizontalis

English:
Cypress

Maltese:
Ċipress Kannella

Habitat/Distribution:
Cultivated on different sites, common

Comments:
Cones all year round. Cylindrical type (var. pyramidalis) cultivated as ornament

CUPRESSACEAE

007

Tetraclinis articulata

English:
Sandarac

Maltese:
Siġra ta' l-Għargħar

Habitat/Distribution:
Maquis, few wild specimens

Comments:
The national tree, sometimes cultivated. Small cones with 4 scales

LAURACEAE

Magnoliopsida

Family:
Lauraceae

008

Laurus nobilis

English:
Bay Laurel

Maltese:
Siġra tar-Rand

Habitat/Distribution:
Valleys, maquis, rare

Flowering time:
Spring

Comments:
Indigenous evergreen tree, black fruits. Aromatic leaves, ornamental plant

ARISTOLOCHIACEAE

Family:
Aristolochiaceae

009

Aristolochia longa

English:
Green-flowered Birthworth

Habitat/Distribution:
Rocky sites, rare

Flowering time:
Spring

Comments:
Flower 3 cm

ALISMATACEAE

Liliopsida

Family:
Alismataceae

010

Damasonium bourgaei (Damasonium alisma ssp. burgaei)

English:
Mediterranean Starfruit

Maltese:
Damażonju

Habitat/Distribution:
Garigue, rainwater pools, rare

Flowering time:
Late spring

010

ARACEAE

Family:
Araceae

011

Arisarum vulgare

English:
Small Lords-and-Ladies, Friar's Cowl

Maltese:
Garni tal-Pipi

Habitat/Distribution:
Damp, fertile sheltered sites, common

Flowering time:
Autumn-spring

ARACEAE

012-014

Arum italicum

English:
Italian Lords-and-Ladies

Maltese:
Garni

Habitat/Distribution:
Valleys, frequent

Flowering time:
Winter-spring

Comments:
Stalk with sterile, male and female flowers enveloped in a bract (bract was removed in fig. 013): insect-trap for pollination! Poisoness berries (see fig. 014)

012

ARACEAE – COLCHICACEAE – IRIDACEAE

013

014

Family:
Colchicaceae

015

Colchicum cupani

English:
Mediterranean Meadow Saffron

Maltese:
Busieq

Habitat/Distribution:
Garigue, rocky steppes, common

Flowering time:
Autumn

Comments:
Small leaves developed in winter. Flower 1-3 cm

015

016

Family:
Iridaceae

016

Freesia refracta

English:
Common Freesea

Maltese:
Friżja

Habitat/Distribution:
Maquis, valleys, frequent

Flowering time:
Spring

Comments:
Introduced from S.Africa, cultivated for ornament, naturalised

25

IRIDACEAE

017

Crocósmia x aurea

English:
Montbretia

Maltese:
Monbrezja

Habitat/Distribution:
Disturbed grounds, frequent-common

Flowering time:
Spring

Comments:
Introduced from S.Africa, cultivated for ornament, naturalised

018

Gladiolus italicus (Gladiolus segetum)

English:
Field Gladiolus

Maltese:
Ħabb il-Qamħ tar-Raba'

Habitat/Distribution:
Close to cultivated fields, common

Flowering time:
Spring

Comments:
Cultivated as ornament

017

IRIDACEAE

018

019

019

Gynandriris sisyrinchium (Iris sisyrinchium)

English:
Barbary Nut Iris

Maltese:
Fjurdulis Salvaġġ

Habitat/Distribution:
Garigue, rocky sites, frequent

Flowering time:
Spring

Comments:
Flowers open in afternoon, 3-5 cm

020

020

Romulea ramiflora

English:
Sand-crocus

Maltese:
Żagħfran tal-Blat

Habitat/Distribution:
Rocky sites, common

Flowering time:
Spring

Comments:
Flower 2-4 cm

SMILACACEAE – ASPHODELACEAE

Family:
Smilacaceae

021, 022

Smilax aspera

English:
Common Smilax

Maltese:
Salsa Pajżana

Habitat/Distribution:
Valleys, maquis, common

Flowering time:
Autumn

Comments:
Climbing prickly plants

021

022

Family:
Asphodelaceae

023

Asphodelus aestivus

English:
Branched Asphodel

Maltese:
Berwieq

Habitat/Distribution:
Steppes, garigue, common

Flowering time:
Spring

Comments: Bluish-green grass-like leaves

023

024

ASPHODELACEAE – ORCHIDACEAE

024

Asphodelus fistulosus

English:
Pink Asphodel

Maltese:
Berwieq Żghir

Habitat/Distribution:
Rocky places, rare

Flowering time:
Spring

Comments:
Smaller in size as Branched Asphodel

Family:
Orchidaceae

025

Orchis collina (Orchis saccata)

English:
Fan-lipped Orchid

Maltese:
Orkida Ħamra

Habitat/Distribution:
Fallow fields, garigue, frequent

Flowering time:
Winter-spring

Comments:
Height 10-20 cm. All orchids are worldwide protected

ORCHIDACEAE

026

Orchis coriophora ssp. fragrans

English:
Scented Bug Orchid

Maltese:
Orkida Tfuh

Habitat/Distribution:
Garigue, rocky sites, rare-frequent

Flowering time:
Spring

Comments:
1 orchid-flower can produce up to 1 million seeds

027

Orchis conica (Orchis lactea)

English:
Milky Orchid

Maltese:
Orkida tat-Tikek

Habitat/Distribution:
Garigue, rocky sites, frequent

Flowering time:
Spring

Comments:
Orchid-flowers are variable

028

Anacamptis pyramidalis

English:
Common Pyramidal Orchid

Maltese:
Orkida Piramidali

Habitat/Distribution:
Garigue, maquis, common

026

027

028

029

ORCHIDACEAE

Flowering time:
Late spring

Comments:
All orchids rely on specific fungus for their germination and development

029

Anacamptis urvilleana

English:
Maltese Pyramidal Orchid

Maltese:
Orkida Piramidali ta' Malta

Habitat/Distribution:
Garigue, maquis, frequent

Flowering time:
Early spring

Comments:
Probably endemic to Malta/Sicily, this orchid is rather unknown

030

Ophrys fusca

English:
Brown Orchid

Maltese:
Dubbiena

Habitat/Distribution:
Garigue, maquis, steppes, common

Flowering time:
Winter-spring

Comments:
Flower 2 cm

ORCHIDACEAE

031

**Ophrys speculum
(Ophrys vernixia,
Ophrys ciliata)**

English:
Mirror Orchid

Maltese:
Dubbiena Kahla

Habitat/Distribution:
Garigue, rare

Flowering time:
Spring

Comments:
Labellum has metallic color

032

Ophrys bertolonii

English:
Bertoloni's Orchid

Maltese:
Dubbiena ta' Bertoloni

Habitat/Distribution:
Rocky steppes, very rare

Flowering time:
Spring

Comments:
Ophrys flowers use form, color and smell to attract pollinators

031

032

033

ORCHIDACEAE – ALLIACEAE

033

Ophrys bombyliflora

English:
Bumble Bee Orchid

Maltese:
Nahla

Habitat/Distribution:
Garigue, rocky steppe, frequent

Flowering time:
Spring

Comments:
Flower 1-1.5 cm. The most common Ophrys species of Malta

Family:
Alliaceae

034

Allium lojaconoi

English:
Maltese Dwarf Garlic

Maltese:
Tewm Irqiq ta' Malta

Habitat/Distribution:
Rocky sites of valleys, rare. Endemic!

Flowering time:
Summer

Comments:
Very delicate plant, height 4-8 cm

ALLIACEAE

035

Allium commutatum

English:
Wild Leek

Maltese:
Kurrat Salvaġġ

Habitat/Distribution:
Rocky steppes, frequent

Flowering time:
Spring-summer

Comments:
Many similar species occur

036

Allium subhirsutum

English:
Hairy Garlic

Maltese:
Tewm Muswaf

Habitat/Distribution:
Rocky sites of valleys, maquis, common

Flowering time:
Late spring

Comments:
Almost 120 Allium species are distributed in Europe, including the cultivated Leeks Allium porrum and Allium cepa

035

036

037

AMARYLLIDACEAE

Family:
Amaryllidaceae

037

Narcissus serotinus

English:
Autumn Narcissus

Maltese:
Narċis Imwahhar

Habitat/Distribution:
Rocky places, garigue, common

Flowering time:
Autumn

Comments:
Occurrence even on disturbed places, height 5-20 cm

038

Narcissus tazetta

English:
French Daffodil

Maltese:
Ranġis, Narċis

Habitat/Distribution:
Valleys, neglected fields, common

Flowering time:
Winter

Comments:
Extremely sweet smell

AMARYLLIDACEAE

039

Pancratium maritimum

English:
Sea Daffodil

Maltese:
Pankrazju, Narċis il-Baħar

Habitat/Distribution:
Sandy beaches, rare

Flowering time:
Late summer

DIOSCOREACEAE

Family:
Dioscoreaceae

040

Tamus communis

English:
Black Bryony

Maltese:
Brijonja Sewda

Habitat/Distribution:
Between rocks, valleys, rare

Flowering time:
Early summer

Comments:
Small flowers are greenish-yellow

HYACINTHACEAE

Family:
Hyacinthaceae

041

Muscari comosum

English:
Tassel Hyacinth

Maltese:
Basal il-Ħnieżer

Habitat/Distribution:
Rocky steppes, neglected fields, common

Flowering time:
Spring

Comments:
Sterile head flowers

042

Muscari parviflorum

English:
Autumn Grape Hyacinth (Lesser Grape Hyacinth)

Maltese:
Ġjaċint tal-Ħarifa

Habitat/Distribution:
Rocky steppes, garigue, frequent

Flowering time:
Autumn

Comments:
Small plants, flower 0.5cm, none or few sterile head flowers

043

Scilla sicula

English:
Sicilian Squill

Maltese:
Għansar Ikħal

HYACINTHACEAE

Habitat/Distribution:
Valleys, garigue, maquis, rare. Endemic to Malta, Sicily and Calabria

Flowering time:
Spring

Comments:
Cultivated plants of Scilla peruviana have deep blue flowers

044

Scilla autumnalis

English:
Autumn Squill

Maltese:
Ghansar tal-Harifa

Habitat/Distribution:
Rocky places, garigue, common

Flowering time:
Autumn

Comments:
Height 3-10 cm

045

Urginea pancration

English:
Sea Squill

Maltese:
Ghansar

Habitat/Distribution:
Rocky ground, common

Flowering time:
Late summer

Comments:
Plants from Malta/Italy recently separated from the species Urginea maritima, leaves of both (see fig. 193) occur in autumn

ASPARAGACEAE – AGAVACEAE

Family:
Asparagaceae

046

Asparagus aphyllus

English:
Southern Spiny Asparagus

Maltese:
Spraġ Xewwieki

Habitat/Distribution:
Valleys, maquis, fallow fields, common

Flowering time:
Autumn

Comments:
Shrub-like plants, strongly spiny, black berries. Flower 1-1.5 cm

046

047

AGAVACEAE

Family:
Agavaceae

047, 048

Agave americana

English:
American Agave

Maltese:
Sabbara ta' l-Amerika

Habitat/Distribution:
Rocky places near the coast, common

Flowering time:
Early summer

Comments:
Naturalised, grows worldwide under subtropical arid conditions

CYPERACEAE – JUNCACEAE

Family:
Cyperaceae

049

Holoschoenus vulgaris (Scirpus holoschoenus)

English:
Round-headed Club-rush

Maltese:
Simar tal-Boċċi

Habitat/Distribution:
Watercourses, common

Flowering time:
Early summer

050

Cyperus capitatus

English:
Sand Galingale

Maltese:
Bordi tal-Ramel

Habitat/Distribution:
Coast, sandy places, rare

Flowering time:
Early summer

Family:
Juncaceae

051

Juncus sorrentini

English:
Toad Rush

Maltese:
Simar Żgħir

Habitat/Distribution:
Rocky places, rare

JUNCACEAE – POACEAE

Flowering time:
Early summer

Comments:
Height 10 cm. Species belong to the group of Juncus bufonius

Family:
Poaceae

052

Sporolobus pungens (Sporolobus arenarius)

English:
Sand Dropseed

Maltese:
Niġem tar-Ramel

Habitat/Distribution:
Coast, sandy places, rare

Flowering time:
Summer

053

Hordeum leporinum

English:
Hare's-tail Berley

Maltese:
Nixxief Komuni

Habitat/Distribution:
Disturbed sites, common

Flowering time:
Spring

Comments:
Species was separated from the group of Hordeum murinum

050

051

052

053

43

POACEAE

054

Lygeum spartum

English:
Esparto Grass

Maltese:
Halfa

Habitat/Distribution:
Clay steppes, frequent

Flowering time:
Spring

Comments:
Eye-catching is an inflorescence-sheath

055

Arundo donax

English:
Great Reed

Maltese:
Qasba Kbira

Habitat/Distribution:
Watercourses of valleys, wet sites on slopes, common

Flowering time:
Autumn

Comments:
Plants are used for shelter, embouchures for clarinet, manufactore of baskets

POACEAE

056

Avena sterilis

English:
Animated Oat

Maltese:
Harfur Kbir

Habitat/Distribution:
Disturbed places, common

Flowering time:
Spring

057

Aegilops geniculata (Aegilops ovata)

English:
Goat Grass

Maltese:
Brimba

Habitat/Distribution:
Steppes, frequent

Flowering time:
Spring

058

Cynodon dactylon

English:
Bermuda Grass

Maltese:
Niġem

Habitat/Distribution:
Disturbed places, common

Flowering time:
Spring-autumn

POACEAE

059

Polypogon monspeliensis

English:
Annual Beard-grass

Maltese:
Denb il-Liebru Kbir

Habitat/Distribution:
Valleys, frequent

Flowering time:
Early summer

POACEAE

060

Lagurus ovatus

English:
Hare's-tail Grass

Maltese:
Denb il-Fenek

Habitat/Distribution:
Garigue, steppes, disturbed places, frequent

Flowering time:
Spring

Comments:
Sometimes cultivated as ornament

POACEAE – RANUNCULACEAE

061

Hyparrhenia hirta

English:
Beard-grass

Maltese:
Barrum Vjola

Habitat/Distribution:
Open habitats, common

Flowering time:
Spring-autumn

Rosopsida

Family:
Ranunculaceae

062

Clematis cirrhosa

English:
Evergreen Traveller's Joy

Maltese:
Kiesha

Habitat/Distribution:
Maquis, scarce

Flowering time:
Autumn

Comments:
Climbing shrub

063

Anemone coronaria

English:
Crown Anemone

Maltese:
Kahwiela

Habitat/Distribution:
Maquis, valleys

Flowering time:
Winter-spring

48

RANUNCULACEAE

064

Ranunculus ficaria

English:
Lesser Celandine

Maltese:
Fomm l-Gheliem

Habitat/Distribution:
Valleys, maquis, rare-frequent

Flowering time:
Spring

Comments:
This subspecies with large flowers is distributed in S.Europe

065

Ranunculus bullatus

English:
Autumn Buttercup

Maltese:
Ċfolloq

Habitat/Distribution:
Almost everywhere

Flowering time:
Autumn

066

Ranunculus muricatus

English:
Scilly Buttercup

Maltese:
Ċfolloq

Habitat/Distribution:
Valleys, common

Flowering time:
Spring-summer

Comments:
Height 10-40 cm

RANUNCULACEAE – PAPAVERACEAE

067

Ranunculus saniculaefolius

English:
Sanicle-leaved Water Crowfoot

Maltese:
Ċfolloq ta' l-Ilma

Habitat/Distribution:
Rock pools in Garigue, rare

Flowering time:
Winter-spring

Comments:
Two kinds of leaves

068

Ranunculus trichophyllus

English:
Three-leaved Water Crowfoot

Maltese:
Ċfolloq tal-Wied

Habitat/Distribution:
Watercourses in valleys, rare

Flowering time:
Winter-summer

Comments:
Flower 1 cm

067

068

069

PAPAVERACEAE

Family:
Papaveraceae

069, 070

Glaucium flavum

English:
Yellow Horned-poppy

Maltese:
Pepprin Isfar

Habitat/Distribution:
Sandy, disturbed sites, common

Flowering time:
Spring-autumn

070

51

TAMARICACEAE – FRANKENIACEAE

Family:
Tamaricaceae

071

Tamarix africana

English:
African Tamarisk

Maltese:
Siġra tal-Bruk

Habitat/Distribution:
Coast, common

Flowering time:
Spring-summer

Comments:
Indigenous and introduced plants

071

Family:
Frankeniaceae

072

Frankenia hirsuta (Frankenia laevis var. hirsuta)

English:
Sea-heath

Maltese:
Erba Franka

Habitat/Distribution:
Sandy, near the sea, frequent

Flowering time:
Spring

Comments:
Hairy, flower 1 cm

072

CARYOPHYLLACEAE

Family:
Caryophyllaceae

073

Silene colorata

English:
Red Campion

Maltese:
Lsien l-Għasfur

Habitat/Distribution:
Everywhere

Flowering time:
Winter-spring

CHENOPODIACEAE

Family:
Chenopodiaceae

074, 075

Arthrocnemum macrostachyum (Arthrocnemum glaucum)

English:
Shrubby Glasswort

Maltese:
Almeridja tal-Blat

Habitat/Distribution:
Rocky coasts, common

Flowering time:
Late spring

Comments:
Extremely reduced flowers (fig. 075)

076

Salsola kali

English:
Prickly Saltwort

Maltese:
Ħaxixa ta' l-Irmied Xewwikija

Habitat/Distribution:
Shores, rare

Flowering time:
Summer

Comments:
Stem 20-80 cm

CHENOPODIACEAE

077

Salsola soda

English:
Smooth-leaved Saltwort

Maltese:
Ħaxixa ta' l-Irmied

Habitat/Distribution:
Shores, common

Flowering time:
Summer

078

Cremnophyton lanfrancoi

English:
Maltese Cliff-orache

Maltese:
Bjanka ta' l-Irdum

Habitat/Distribution:
Coast, rare. Endemic!

Flowering time:
Summer

Comments:
Named after the Maltese botanist Edwin Lanfranco

CHENOPODIACEAE

079, 080

Atriplex halimus

English:
Shrubby Orache

Maltese:
Bjanka

Habitat/Distribution:
Coast, common

Flowering time:
Summer

Comments:
Cultivated as hedges

081

Darniella melitensis

English:
Maltese Salt-tree

Maltese:
Xebb

Habitat/Distribution:
Coast, common. Endemic!

Flowering time:
Summer

CHENOPODIACEAE – CACTACEAE – NYCTAGINACEAE

080

081

Family:
Cactaceae

082

Opuntia ficus-indica

English:
Prickly Pear

Maltese:
Bajtar tax-Xewk

Habitat/Distribution:
Naturalised at different sites, common

Flowering time:
Summer

Comments:
Fodder plant, fruits edible. Leaves reduced to spines

082

083

Family:
Nyctaginaceae

083

Mirabilis jalapa

English:
Marvel-of-Peru

Maltese:
Ħommejr

Habitat/Distribution:
Disturbed sites, frequent

Flowering time:
Summer

Comments:
Flower 2-4 cm, sometimes white or yellow. Cultivated for ornament

57

POLYGONACEAE

Family:
Polygonaceae

084

Rumex bucephalophorus

English:
Red Dock

Maltese:
Qarsajja

Habitat/Distribution:
Garigue, sandy sites, common

Flowering time:
Spring

Comments:
Height 10-40 cm

PLUMBAGINACEAE

Family:
Plumbaginaceae

085

Limonium virgatum

English:
Seaside Sea-lavender

Maltese:
Limonju tal-Baħar

Habitat/Distribution:
Coast, frequent

Flowering time:
Summer-autumn

Comments:
Some very similar species are endemic

AIZOACEAE – GERANIACEAE

Family:
Aizoaceae

086

Mesembryanthemum nodiflorum

English:
Lesser Crystal-plant

Maltese:
Kristallina Komuni

Habitat/Distribution:
Coast, frequent

Flowering time:
Summer

Comments:
Creeping, small succulent leaves

Family:
Geraniaceae

087

Erodium malacoides

English:
Glandular Storksbill

Maltese:
Moxt

Habitat/Distribution:
Almost everywhere

Flowering time:
Spring

088

Erodium moschatum

English:
Musk Storksbill

Maltese:
Ħaxixa tal-Misk

GERANIACEAE – ZYGOPHYLLACEAE

Habitat/Distribution:
Various sites, common

Flowering time:
Spring

Comments:
Height 5-50 cm

Family:
Zygophyllaceae

089

Fagonia cretica

English:
Fagonia

Maltese:
Fagonja

Habitat/Distribution:
Clay sites, rare

Flowering time:
Spring

Comments:
Often climbing between shrubs

090

Tribulus terrestris

English:
Maltese Cross

Maltese:
Ghatba

Habitat/Distribution:
Valletta, anthropomorphic sites, rare

Flowering time:
Late summer

Comments:
Almost worldwide in suitable areas. Fruit resembles the Maltese cross

61

CRASSULACEAE – CLUSIACEAE

Family:
Crassulaceae

091

Sedum sediforme

English:
Mediterranean Stonecrop

Maltese:
Sedum

Habitat/Distribution:
Garigue, rocky sites, frequent

Flowering time:
Early summer

092

Sedum caeruleum

English:
Blue Stonecrop

Maltese:
Beżżul il-Baqra

Habitat/Distribution:
Rocky sites, desiccated freshwater pools, frequent

Flowering time:
Spring

Comments:
Height 3-15 cm

091

092

093

62

CLUSIACEAE

Family:
Clusiaceae

093

Hypericum aegyptiacum (Triadenia aegyptica)

English:
Egyptian St. John's-wort

Maltese:
Fexfiex ta' l-Irdum

Habitat/Distribution:
Garigue, cliffs, rocky sites, frequent

Flowering time:
Winter-late spring

094

Hypericum pubescens

English:
Pubescent St. John's-wort

Maltese:
Fexfiex Sufi

Habitat/Distribution:
Almost everywhere

Flowering time:
Spring-summer

Comments:
Hairy plant

CLUSIACEAE – LINACEAE

095

Hypericum triquetrifolium (Hypericum crispum)

English:
Crisped St. John's-wort

Maltese:
Fexfiex tar-Raba'

Habitat/Distribution:
Disturbed sites, frequent

Flowering time:
Summer

Family:
Linaceae

096

Linum strictum

English:
Upright Yellow Flax

Maltese:
Kittien Isfar

Habitat/Distribution:
Garigue, rocky sites, frequent

Flowering time:
Spring

097

Linum trigynum

English:
Southern Flax

Maltese:
Kittien Irqiq

Habitat/Distribution:
Garigue, rocky sites, frequent

Flowering time:
Spring

EUPHORBIACEAE

Family:
Euphorbiaceae

098, 099
Mercurialis annua

English:
Annual Mercury

Maltese:
Burikba

Habitat/Distribution:
Disturbed sites, common

Flowering time:
Winter-spring

Comments:
Male plant (fig. 098), female plant (fig. 099)

100
Ricinus communis

English:
Castor Oil Tree

Maltese:
Siġra tar-Riċnu

Habitat/Distribution:
Disturbed sites, valleys, frequent, naturalised. Africa

Flowering time:
All year round, common

Comments:
Cultivated in for oil, medicine and as ornament

101
Euphorbia paralias

English:
Sea Spurge

Maltese:
Tenghud tar-Ramel

Habitat/Distribution:
Sandy beaches, rare

Flowering time:
Summer

65

EUPHORBIACEAE

102, 103

Euphorbia melitensis

English:
Maltese Spurge

Maltese:
Tenghud tax-Xaghri

Habitat/Distribution:
Garigue, frequent.
Endemic!

Flowering time:
Spring

104

Euphorbia pinea

English:
Pine Spurge

Maltese:
Tenghud Komuni

Habitat/Distribution:
Rocky sites, disturbed areas, frequent

Flowering time:
All year round

102

103

104

EUPHORBIACEAE – OXALIDACEAE

105

Euphorbia dendroides

English:
Tree Spurge

Maltese:
Tenghud tas-Siġra

Habitat/Distribution:
Garigue, frequent

Flowering time:
Spring

Comments:
Leaf color from winter to early summer yellow, green, red, brown

Family:
Oxalidaceae

106

Oxalis pes-caprae

English:
Cape Sorrel

Maltese:
Haxixa Ingliża

Habitat/Distribution:
Almost everywhere

Flowering time:
Spring

Comments:
Sterile plants from S. Africa. Common is a double-flowered type (Ingliża Sewda)

MIMOSACEAE – CYNOMORIACEAE – CAESALPINIACEAE

Family:
Mimosaceae

107

Acacia cyanophylla

English:
Blue-leaved Acacia

Maltese:
Gazzija

Habitat/Distribution:
Steppes, maquis, disturbed places, common

Flowering time:
Spring

Comments:
Cultivated but naturalised, lead to ecological problems

107

Family:
Cynomoriaceae

108

Cynomorium coccineum

English:
Malta Fungus, General's root

Maltese:
Għerq Sinjur, Għerq il-Ġeneral

Habitat/Distribution:
Rocky sites, Fungus Rock, Dingli, very rare

Flowering time:
Spring

Comments:
Stem 15-30 cm. Parasitic on halophytic shrubs. Probably formerly

108  109

CYNOMORIACEAE – CAESALPINIACEAE

110

endemic on Fungus Rock. Long thought to be a fungal miracle drug, therfore, introduced to other Mediterranean areas (this picture)

Note:
In plant systematics, the family *Cynomoriaceae* (fig. 108) is placed between the families *Oxalidaceae* (fig. 106) and *Mimosaceae* (fig. 107). It has been placed between *Mimosaceae* (fig. 107) and *Caesalpiniaceae* (fig. 109) for technical reasons.

Family:
Caesalpiniaceae

109-111

Ceratonia siliqua

English:
Carob Tree

Maltese:
Siġra tal-Ħarrub

Habitat/Distribution:
Valleys, common

Flowering time:
Autumn

Comments:
Flowers of a female (fig. 109) and a male tree (fig. 110). Cultivated to harvest fruits (fig.111) for fodder and food industry in S.Europe

111

69

FABACEAE

Family:
Fabaceae

112, 113

Anthyllis hermanniae

English:
Shrubby Kidney Vetch

Maltese:
Ħatba Sewda

Habitat/Distribution:
Garigue, frequent

Flowering time:
Spring

112

113

FABACEAE

114

Anthyllis vulneraria ssp. maura

English:
Common Kidney Vetch

Maltese:
Silla tal-Blat

Habitat/Distribution:
Garigue, frequent

Flowering time:
Spring

114

115

Coronilla valentina var. glauca

English:
Shrubby Crown Vetch

Maltese:
Koronilla

Habitat/Distribution:
Rocky sites of valleys, rare

Flowering time:
Spring

115

71

FABACEAE

116

Coronilla scorpioides

English:
Yellow Crown Vetch

Maltese:
Xeħt l-Imħabba; Morra

Habitat/Distribution:
Open, disturbed areas, rare

Flowering time:
Spring

Comments:
Height 10-40 cm

117

Scorpiurus muricatus

English:
Scorpion-tail Vetch

Maltese:
Widna

Habitat/Distribution:
Garigue, frequent

Flowering time:
Spring
Comments: Flower 1 cm

118

Trifolium stellatum

English:
Star Clover

Maltese:
Xnien ta' l-Istilla

Habitat/Distribution:
Garigue, steppes, frequent

Flowering time:
Spring

FABACEAE

119

Psoralea bituminosa

English:
Pitch Clover

Maltese:
Silla tal-Mogħoż

Habitat/Distribution:
Maquis, garigue, common

Flowering time:
Spring

Comments:
Plants smell like asphalt

120

Ononis natrix ssp. ramosissima

English:
Bushy Restharrow

Maltese:
Broxka ta' Għawdex

Habitat/Distribution:
Garigue, sand dunes, rare

Flowering time:
Spring

FABACEAE

121

Lotus tetragonolobus (Tetragonolobus purpureus)

English: Winged Pea

Maltese: Fiġġiela Ħamra

Habitat/Distribution: Almost everywhere

Flowering time: Spring

Comments: Height 10-30cm

122

Lotus ornithopodioides

English: Common Birdsfoot Trefoil

Maltese: Qrempuċ tal-Mogħoż

Habitat/Distribution: Almost everywhere

Flowering time: Spring

123

Lotus cytisoides

English: Grey Birdsfoot Trefoil

Maltese: Għantux tal-Blat

Habitat/Distribution: Garigue, rocky sites, frequent

Flowering time: Winter-spring

FABACEAE

124

Lotus edulis

English:
Edible Birdsfoot Trefoil

Maltese:
Qrempuċ

Habitat/Distribution:
Almost everywhere

Flowering time:
Spring

125

Lathyrus clymenum

English:
Crimson Pea

Maltese:
Ġilbiena tas-Serp

Habitat/Distribution:
Almost everywhere

Flowering time:
Spring

126

Hedysarum coronarium

English:
Sulla

Maltese:
Silla

Habitat/Distribution:
Disturbed sites, clay, common

Flowering time:
Spring

Comments:
Cultivated for fodder, naturalised

75

FABACEAE – CUCURBITACAE – ROSACEAE

127

Medicago polymorpha

English:
Toothed Medick

Maltese:
Nefel Komuni

Habitat/Distribution:
Almost everywhere

Flowering time:
Winter-spring

Family:
Cucurbitaceae

128

Ecballium elaterium

English:
Squirting Cucumber

Maltese:
Faqqus il-Ħmir

Habitat/Distribution:
Disturbed places, common

Flowering time:
All year round

Comments:
Ripe fruit exploding

Family:
Rosaceae

129

Rubus ulmifolius

English:
Bramble

Maltese:
Għolliq

Habitat/Distribution:
Maquis, valleys, common

Flowering time:
Spring-autumn

Comments:
Fruit edible

127

128

129

130

76

ROSACEAE

130

Crataegus monogyna x azarolus

English: Hawthorn/Azarole

Maltese: Żagħrun/Anżalor

Habitat/Distribution: Maquis, valleys, frequent

Flowering time: Spring

Comments: Hybrid of the 2 species Crataegus monogyna and C. azarolus

131

Rosa sempervirens

English: Evergreen Rose

Maltese: Girlanda tal-Wied

Habitat/Distribution: Maquis, valley sides, rare

Flowering time: Early summer

132

Potentilla reptans

English: Creeping Cinquefoil

Maltese: Frawla Salvaġġa

Habitat/Distribution: Watercourses, frequent

Flowering time: Early summer

Comments: Flower 1.5-2.5 cm

RHAMNACEAE – FAGACEAE

Family:
Rhamnaceae

133

Rhamnus alaternus

English:
Mediterranean Buckthorn

Maltese:
Alaternu

Habitat/Distribution:
Maquis, valleys, rare

Flowering time:
Winter

Comments:
Red berries

134

Rhamnus oleoides

English:
Olive-leaved Buckthorn

Maltese:
Żiju

Habitat/Distribution:
Garigue, Maquis, valley sides, frequent

Flowering time:
Winter

Comments:
Small, white-yellowish flowers

133

134

135

78

FAGACEAE – LYTHRACEAE

Family:
Fagaceae

135

Quercus ilex

English:
Evergreen Oak

Maltese:
Siġra tal-Ballut

Habitat/Distribution:
Indigenous plants are very rare. Meditteranean region

Flowering time:
Spring

Comments:
This hundreds of years old tree is member of an Oak-tree forest of Wardija

Family:
Lythraceae

136

Lythrum junceum

English:
Creeping Loosestrife

Maltese:
Litrum ta' l-Ilma

Habitat/Distribution:
Watercourses, frequent

Flowering time:
Early summer

Comments:
Height 20-70 cm

CAPPARIDACEAE – BRASSICACEAE

Family:
Capparidaceae

137

Capparis orientalis (Capparis spinosa var. inermis, Capparis rupestris)

English:
Caper

Maltese:
Kappara

Habitat/Distribution:
On rocky sites, walls, common

Flowering time:
Summer

Comments:
Edible capers are flower buds. Worldwide, "Kappara" taste best!

137

Family:
Brassicaceae

138

Diplotaxis erucoides

English:
White Mustard

Maltese:
Ġarġir Abjad

Habitat/Distribution:
Disturbed sites, very common

Flowering time:
Autumn-spring

138

139

BRASSICACEAE

139

Diplotaxis tenuifolia

English:
Perennial Wall-rocket

Maltese:
Ġarġir Isfar

Habitat/Distribution:
Disturbed sites, very common

Flowering time:
All year round

Comments:
Leaves of both species can be used for salad

140

Matthiola incana ssp. melitensis

English:
Maltese Stocks

Maltese:
Ġiżi ta' Malta

Habitat/Distribution:
Coast, cliffs, garigue, rare. Endemic!

Flowering time:
Spring

BRASSICACEAE

141

Matthiola tricuspidata

English:
Mediterranean Stocks

Maltese:
Ġiżi tal-Baħar

Habitat/Distribution:
Sandy Coast, rare

Flowering time:
Spring

142

Cakile maritima

English:
Sea Rocket

Maltese:
Kromb il-Baħar

Habitat/Distribution:
Coast, frequent

Flowering time:
Spring-summer

143

Lobularia maritima

English:
Sweet Alison

Maltese:
Buttuniera

Habitat/Distribution:
Almost everywhere

Flowering time:
All year round

Comments:
Height 10-40 cm

RESEDACEAE

Family:
Resedaceae

144

Reseda alba

English:
White Mignonette

Maltese:
Denb il-Ħaruf

Habitat/Distribution:
Disturbed sites, common

Flowering time:
Spring

CISTACEAE – MALVACEAE

Family:
Cistaceae

145

Fumana thymifolia

English:
Thyme-leaved Sun-rose

Maltese:
Ċistu Żghir

Habitat/Distribution:
Garigue, frequent

Flowering time:
Spring

146

Cistus creticus (Cistus incanus ssp. creticus)

English:
Hoary Rock-rose

Maltese:
Ċistu Roża

Habitat/Distribution:
Garigue, maquis, rare

Flowering time:
Spring

Comments:
Flower 5-6 cm

Family:
Malvaceae

147

Malva sylvestris

English:
Common Mallow

Maltese:
Ħobbejża Komuni

Habitat/Distribution:
Garigue, disturbed sites, common

145

146

MALVACEAE – ANACARDIACEAE

Flowering time:
Spring

Comments:
Height 2-100 cm

148

Lavatera arborea

English:
Tree Mallow

Maltese:
Ħobbejża tas-Siġra

Habitat/Distribution:
Disturbed sites, frequent

Flowering time:
Spring-summer

Family:
Anacardiaceae

149

Pistacia lentiscus

English:
Lentisk

Maltese:
Deru

Habitat/Distribution:
Maquis, valleys

Flowering time:
Winter-spring

Comments:
Flowers small, brownish (females)

RUTACEAE – ERICACEAE

Family:
Rutaceae

150

Ruta chalepensis

English:
Fringed Rue

Maltese:
Fejġel

Habitat/Distribution:
Garigue, common

Flowering time:
Spring

Comments:
Aromatic smell when touching

150

151

Family:
Ericaceae

151, 152

Erica multiflora

English:
Mediterranean Heath

Maltese:
Erika

Habitat/Distribution:
Garigue, maquis, common

Flowering time:
Spring

152

RUBIACEAE

Family:
Rubiaceae

153

Putoria calabrica

English:
Stinking Madder

Maltese:
Robbja, Alizzari

Habitat/Distribution:
Garigue, rare

Flowering time:
Spring

RUBIACEAE

154

Crucianella rupestris

English:
Rock Crosswort

Maltese:
Kruċanella

Habitat/Distribution:
Cliffs, seaside rocks, frequent

Flowering time:
Late spring

Comments:
Flower 1 cm

155

Asperula cynanchica

English:
Squinancywort

Habitat/Distribution:
Garigue, frequent

Flowering time:
Late spring

Comments:
Flower 1.5 cm

156

Valantia muralis

English:
Hairy Valantia

Habitat/Distribution:
Garigue, common

Flowering time:
Spring

Comments:
Flowers 1-1.5 mm

RUBIACEAE – ASCLEPIADACEAE

157

Rubia peregrina

English:
Wild Madder

Maltese:
Robbja Salvaġġa

Habitat/Distribution:
Maquis, valleys, frequent

Flowering time:
Spring

Comments:
Climbing, small flowers greenish

Family:
Asclepiadaceae

158, 159

Periploca angustifolia

English:
Wolfbane

Maltese:
Siġra tal-Ħarir

Habitat/Distribution:
Garigue, maquis, frequent

Flowering time:
Autumn-spring

157

158

159

GENTIANACEAE – LAMIACEAE

Family:
Gentianaceae

160

Blackstonia perfoliata

English:
Yellow-wort

Maltese:
Ċentawrja Safra

Habitat/Distribution:
Rocky sites, frequent

Flowering time:
Late spring

161

Centaurium erythraea

English:
Common Centaury

Maltese:
Ċentawrja Kbira

Habitat/Distribution:
Rocky sites, frequent

Flowering time:
Late spring

160

161

162

90

LAMIACEAE

Family:
Lamiaceae

162, 163

Teucrium fruticans

English:
Olive-leaved Germander

Maltese:
Żebbuġija

Habitat/Distribution:
Garigue, maquis, common

Flowering time:
Winter-summer

LAMIACEAE

164

Teucrium flavum

English:
Yellow Germander

Maltese:
Borgħom Komuni

Habitat/Distribution:
Garigue, maquis, common

Flowering time:
Spring-summer

164

165

Ajuga iva

English:
Southern Bugle

Maltese:
Xantkura

Habitat/Distribution:
Garigue, frequent

Flowering time:
Late spring

Comments:
Plants with yellow flowers seem to belong to the same species. They prefer more disturbed grounds

165

LAMIACEAE

166

Phlomis fruticosa

English:
Great Sage

Maltese:
Salvja tal-Madonna

Habitat/Distribution:
Areas with coralline rocks, frequent

Flowering time:
Late spring

Comments:
Ornamental plant in Europe

LAMIACEAE

167, 168

Prasium majus

English:
White Hedge-nettle

Maltese:
Te Sqalli

Habitat/Distribution:
Garigue, valley-slopes, common

Flowering time:
Spring

169, 170

Thymbra capitata (Thymus capitatus, Coridothymus capitatus)

English:
Mediterranean Thyme

Maltese:
Saghtar

Habitat/Distribution:
Garigue, common

Flowering time:
Late spring

Comments:
Bee plant (Maltese Thyme Honey!), beautiful aromatic smell when touching

LAMIACEAE

171

Micromeria microphylla (Satureja microphylla)

English:
Maltese Savory

Maltese:
Xpakkapietra

Habitat/Distribution:
Rocky sites, common

Flowering time:
Winter-summer

172

Salvia verbenaca

English:
Wild Clary

Maltese:
Salvja Salvaġġa

Habitat/Distribution:
Different areas, frequent-common

Flowering time:
Autum-spring

LAMIACEAE – VERBENACEAE

173

Mentha pulegium

English:
Pennyroyal

Maltese:
Plejju

Habitat/Distribution:
Valley-bottoms, frequent-common

Flowering time:
Spring-summer

Family:
Verbenaceae

174

Vitex agnus-castus

English:
Chaste Tree

Maltese:
Siġra tal-Virgi, Bżar tal-Patrijiet

Habitat/Distribution:
Valley-bottoms, rare

Flowering time:
Late summer

175

Verbena officinalis

English:
Vervain

Maltese:
Buqexrem

Habitat/Distribution:
Valley-bottoms, common

Flowering time:
Spring-winter

Comments:
Height 40-100 cm

VERBENACEAE – ACANTHACEAE – SCROPHULARIACEAE

174

175

176

177

Family:
Acanthaceae

176

Acanthus mollis

English:
Bear's Breeches

Maltese:
Ħannewija

Habitat/Distribution:
Maquis, valleys, common

Flowering time:
Late spring

Family:
Scrophulariaceae

177

Verbascum sinuatum

English:
Wavy-leaved Mullein

Maltese:
Xatbet l-Andar

Habitat/Distribution:
Disturbed sites, common

Flowering time:
Spring-summer

SCROPHULARIACEAE

178

Bellardia trixago

English:
Bellardia

Maltese:
Perlina Bajda

Habitat/Distribution:
Rocky sites, frequent

Flowering time:
Spring

Comments:
Parasites, penetrate with small haustoria roots of neighboured plants to get water and minerals

179

Antirrhinum tortuosum (Antirrhinum majus ssp. tortuosum)

English:
Greater Snapdragon

Maltese:
Papoċċi Ħamra

Habitat/Distribution:
Rocky sites, common

Flowering time:
All year round

180

Antirrhinum siculum

English:
Sicilian Snapdragon

Maltese:
Papoċċi Bajda

Habitat/Distribution:
Rocky sites, common

Flowering time:
All year round

Comments:
Endemic to Malta/Sicily, now naturalised in the Meditteranean region

SCROPHULARIACEAE

181

Kickxia commutata

English:
Mediterranean Fluellen

Maltese:
Xatbet l-Art Vjola

Habitat/Distribution:
Disturbed sites, valleys, frequent

Flowering time:
Spring-summer

Comments:
Flowers 1-2 cm

182

Veronica cymbalaria

English:
Pale Speedwell

Maltese:
Veronika

Habitat/Distribution:
Disturbed, rocky sites, frequent

Flowering time:
Spring

Comments:
Height 10 cm

183

Veronica anagallis-aquatica

English:
Blue Water Speedwell

Maltese:
Veronika ta' l-Ilma

Habitat/Distribution:
valley-bottoms, frequent

Flowering time:
Spring

Comments:
Height 30-50 cm

OROBANCHACEAE – PLANTAGINACEAE

Family:
Orobanchaceae

184

Orobanche muteli

English:
Dwarf Broomrape

Maltese:
Budebbus ta' l-Ingliża

Habitat/Distribution:
Almost everywhere

Flowering time:
Spring

Comments:
Achlorophyllous parasite, host usually Cape Sorrel. Flowers whitish-yellow or bluish-purple. Could also be named as Orobanche ramosa f. melitensis

184

Family:
Plantaginaceae

185

Plantago serraria

English:
Toothed Plantain

Maltese:
Biżbula tas-Snien

Habitat/Distribution:
Disturbed sites, pathways, frequent-common

Flowering time:
Spring-sommer

Comments:
Inflorescence similar to that shown in fig. 188

185

PLANTAGINACEAE

186

Plantago afra

English:
Glandular Plantain

Maltese:
Żerrigħ il-Brigħet

Habitat/Distribution:
Maquis, rocky sites, rare-frequent

Flowering time:
Spring

187

Plantago lagopus

English:
Hare's-foot Plantain

Maltese:
Biżbula Komuni

Habitat/Distribution:
Almost everywhere

Flowering time:
Spring

188

Plantago coronopus

English:
Buck's-horn Plantain

Maltese:
Biżbula tal-Baħar

Habitat/Distribution:
Coast, pathways, frequent-common

Flowering time:
Spring-sommer

CONVOLVULACEAE

Family:
Convolvulaceae

189

Convolvulus althaeoides

English:
Mallow Bindweed

Maltese:
Leblieb tax-Xaghri

Habitat/Distribution:
Disturbed sites, common

Flowering time:
Late spring

189

190

Convolvulus elegantissimus

English:
Slender Bindweed

Maltese:
Leblieb tax-Xaghri Ċar

Habitat/Distribution:
Garigue, maquis, common

Flowering time:
Spring

190

CONVOLVULACEAE

191

Convolvulus oleifolius

English:
Olive-leaved Bindweed

Maltese:
Leblieb tal-Blat

Habitat/Distribution:
Garigue, rocky sites, common

Flowering time:
Spring

192

Convolvulus pentapetaloides

English:
Blue Bindweed

Maltese:
Leblieb tal-Werqa Tleqq

Habitat/Distribution:
Garigue, frequent

Flowering time:
Spring

Comments:
Flower 1 cm

CUSCUTACEAE

Family:
Cuscutaceae

193

Cuscuta epithymum

English:
Dodder

Maltese:
Pittma

Habitat/Distribution:
Garigue, common

Flowering time:
Spring-summer

Comments:
Epiphytic parasite lacking green leaves and roots. Preferred hosts on Malta: Mediterranean Thyme, Maltese Spurge, Sea Squill (here)

BORAGINACEAE

Family:
Boraginaceae

194

Echium italicum

English:
Pale Bugloss

Maltese:
Lsien il-Fart Abjad

Habitat/Distribution:
Steppes, disturbed sites, frequent

Flowering time:
Spring

BORAGINACEAE

BORAGINACEAE

195, 196

Borago officinalis

English:
Borage

Maltese:
Fidloqqom

Habitat/Distribution:
Disturbed sites, common

Flowering time:
Spring

BORAGINACEAE

197

Echium parviflorum

English:
Small-flowered Bugloss

Maltese:
Lsien il-Fart Żghir

Habitat/Distribution:
Almost everywhere

Flowering time:
Spring

198

Echium plantagineum

English:
Plantain Bugloss

Maltese:
Lsien il-Fart

Habitat/Distribution:
Coast, sandy pathways, rare

Flowering time:
Spring

199

Cerinthe major

English:
Honeywort

Maltese:
Qniepen

Habitat/Distribution:
Maquis, disturbed sites, common

Flowering time:
Spring

SOLANACEAE

Family:
Solanaceae

200

Hyoscyamus albus

English:
White Henbane

Maltese:
Mammażejża

Habitat/Distribution:
Disturbed sites, common

Flowering time:
All year round

Comments:
Poisonous!

201

Nicotiana glauca

English:
Shrub Tobacco

Maltese:
Tabakk tas-Swar

Habitat/Distribution:
Anthropomorphic sites, common

Flowering time:
Spring-autumn

Comments:
Introduced from S.America

109

APIACEAE

Family:
Apiaceae

202

Smyrnium olusatrum

English:
Alexanders

Maltese:
Karfus il-Ħmir

Habitat/Distribution:
Disturbed habitats, common

Flowering time:
Winter-spring

APIACEAE

203

Daucus carota

English:
Wild Carrot

Maltese:
Zunnarija Salvaġġa

Habitat/Distribution:
Almost everywhere

Flowering time:
Spring-summer

Comments:
Several varieties with different habitats

203

204

Crithmum maritimum

English:
Sea Samphire

Maltese:
Bużbież il-Baħar

Habitat/Distribution:
Coast, rocks, common

Flowering time:
Summer

204

APIACEAE

205

Scandix pecten-veneris

English:
Shepherd's needle

Maltese:
Maxxita

Habitat/Distribution:
Disturbed sites, frequent

Flowering time:
Spring

Comments:
German name "Venuskamm" (Comb of Venus) refers also to the 5 cm long fruit

206

Tordylium apulum

English:
Mediterranean Hartwort

Maltese:
Haxixet it-Trierah

Habitat/Distribution:
Disturbed sites, frequent

Flowering time:
Spring

207

Apium nodiflorum

English:
Procumbent Marshwort

Maltese:
Karfus tal-ilma

Habitat/Distribution:
Valley-bottom, frequent

Flowering time:
Summer

APIACEAE

208

Ferula communis

English:
Giant Fennel

Maltese:
Ferla

Habitat/Distribution:
Garigue, maquis, common

Flowering time:
Spring

Comments:
Leaves of Foeniculum vulgare, Fennel (Bużbież), growing everywhere, are smaller divided, smell

APIACEAE – DIPSACACEAE

209

Eryngium maritimum

English:
Sea Holly

Maltese:
Xewk tar-Ramel

Habitat/Distribution:
Beaches, rare

Flowering time:
Summer-autumn

Family:
Dipsacaceae

210

Scabiosa maritima

English:
Southern Scabious

Maltese:
Skabjoża

Habitat/Distribution:
Garigue, maquis, common

Flowering time:
Spring-summer

Comments:
Inflorescence 1-3 cm

DIPSACACEAE – CAPRIFOLIACEAE – VALERIANACEAE

Family:
Caprifoliaceae

211

Lonicera implexa

English:
Evergreen Honeysuckle

Maltese:
Qarn il-Mogħża

Habitat/Distribution:
Maquis, rocky sites, frequent

Flowering time:
Spring-summer

Comments:
Climbing shrub

210

Family:
Valerianaceae

212

Fedia cornucopiae

English:
Horn-of-plenty

Maltese:
Sieq il-Hamiema

Habitat/Distribution:
Garigue, rocky sites, frequent

Flowering time:
Spring

Comments:
Height 3-25 cm

211

212

115

ASTERACEAE

Family:
Asteraceae

213

Nauplius aquaticus (Asteriscus aquaticus)

English:
Seaside Ox-eye Daisy

Maltese:
Għajn il-Baqra tax-Xatt

Habitat/Distribution:
Garigue, rocky sites, frequent

Flowering time:
Spring

213

214, 215

Inula crithmoides

English:
Golden Samphire

Maltese:
Xorbett

Habitat/Distribution:
Coast, common

Flowering time:
Summer-autumn

214

215

ASTERACEAE

216, 217

**Dittrichia viscosa
(Inula viscosa,
Cupularia viscosa)**

English:
Sticky Fleabane

Maltese:
Tulliera Komuni

Habitat/Distribution:
Disturbed places, common

Flowering time:
Summer

216

217

218

Anthemis urvilleana

English:
Maltese Sea Chamomile

Maltese:
Bebuna tal-Bahar

Habitat/Distribution:
Rocky coast sites, frequent. Endemic!

Flowering time:
Spring

218

ASTERACEAE

219

**Hyoseris radiata
(Hyoseris lucida)**

English:
Perennial Hyoseris

Maltese:
Żigland tal-Pizzi

Habitat/Distribution:
Sheltered rocky sites, frequent

Flowering time:
Winter-spring

220

Hyoseris scabra

English:
Annual Hyoseris

Maltese:
Żigland

Habitat/Distribution:
Garigue, steppes, frequent

Flowering time:
Spring

221, 222

Senecio bicolor

English:
Silvery Ragwort

Maltese:
Kromb il-Baħar Isfar

Habitat/Distribution:
Different sites near the sea, common

Flowering time:
Spring-summer

118

ASTERACEAE

223, 224

Helichrysum melitense (Helichrysum rupestre var. melitense)

English:
Maltese Everlasting

Maltese:
Sempreviva ta' Ghawdex

Habitat/Distribution:
Cliffs of Gozo, rare. Endemic!

Flowering time:
Late spring

223

224

225

Evax pygmaea

English:
Pygmy Cudweed

Maltese:
Evaks

Habitat/Distribution:
Garigue, steppes, common

Flowering time:
Spring

Comments:
Height 1-3 cm

225

119

ASTERACEAE

226

Phagnalon graecum ssp. ginzbergeri

English:
Eastern Phagnalon

Maltese:
Lixka Komuni

Habitat/Distribution:
Garigue, maquis, steppes, common

Flowering time:
Spring

ASTERACEAE

227

Scolymus hispanicus

English:
Sand Oyster Thistle

Maltese:
Xewk Isfar tar-Ramel

Habitat/Distribution:
Beaches, rare

Flowering time:
Spring-summer

228, 229

Aster squamatus

English:
Narrow-leaved Aster

Maltese:
Settembrina Salvaġġa

Habitat/Distribution:
Disturbed, wet places, very common

Flowering time:
Summer-autumn

Comments:
Introduced from S.C.America

227

228

229

ASTERACEAE

230

Picris echioides

English:
Bristly Oxtongue

Habitat/Distribution:
Valley-bottom, rare

Flowering time:
Spring-summer

Comments:
Height 30-150 cm

231

Calendula arvensis

English:
Field Marigold

Maltese:
Suffejra tar-Raba'

Habitat/Distribution:
Almost everywhere

Flowering time:
Winter-spring

Comments:
Annual plants

232

Calendula suffruticosa ssp. fulgida

English:
Shrubby Marigold

Maltese:
Suffejra Kbira

Habitat/Distribution:
Valleys, disturbed sites, frequent

Flowering time:
Winter-summer

Comments:
Perennial plants

ASTERACEAE

233

Pallenis spinosa (Asteriscus spinosus)

English:
Spiny Ox-eye Daisy

Maltese:
Għajn il-Baqra Xewwikija

Habitat/Distribution:
Disturbed grounds, common

Flowering time:
Late Spring

ASTERACEAE

234

Bellis annua

English:
Annual Daisy

Maltese:
Bebuna

Habitat/Distribution:
Garigue, steppes, disturbed ground, common

Flowering time:
Spring

Comments:
Branched

235

Bellis sylvestris

English:
Southern Daisy

Maltese:
Margerita Salvaġġa

Habitat/Distribution:
Maquis, valleys, common

Flowering time:
Autumn-winter

Comments:
Unbranched, similar to Bellis perennis but larger with heights up to 35 cm

236

Senecio leucanthemifolius

English:
Coast Ragwort

Habitat/Distribution:
Garigue, frequent

Flowering time:
Spring

ASTERACEAE

237

Chrysanthemum coronarium

English:
Crown Daisy

Maltese:
Lellux

Habitat/Distribution:
Disturbed ground, common

Flowering time:
Spring

ASTERACEAE

238

Cichorium intybus

English:
Chicory

Maltese:
Ċikwejra

Habitat/Distribution:
Valleys, disturbed sites, frequent

Flowering time:
Spring-summer

Comments:
Height 15-100 cm

239

Cichorium spinosum

English:
Spiny Chicory

Maltese:
Qanfuda

Habitat/Distribution:
Rocky sites, frequent

Flowering time:
Spring

ASTERACEAE

240

Galactites tomentosa

English:
Boar Thistle

Maltese:
Xewk Abjad

Habitat/Distribution:
Disturbed sites, common

Flowering time:
Spring

Comments:
The purple-flowered type is absent

ASTERACEAE

241, 242

Atractylis gummifera

English:
Ground Thistle

Maltese:
Xewk tal-Mixta

Habitat/Distribution:
Garigue, steppes, frequent-common

Flowering time:
Summer-autumn

241

243

Centaurea nicaeensis

English:
Southern Star Thistle

Maltese:
Xewk ta' l-Għotba

Habitat/Distribution:
Steppes, garigue, common

Flowering time:
Spring-summer

Comments:
Stems 30-60 cm, not winged

242

ASTERACEAE

244

Centaurea melitensis

English:
Maltese Star Thistle

Maltese:
Xewk Malti

Habitat/Distribution:
Steppes, garigue, rare

Flowering time:
Spring-summer

Comments:
Winged stems, not endemic as name indicates

245

Chiliadenus bocconei

English:
Maltese Fleabane

Maltese:
Tulliera ta' Malta

Habitat/Distribution:
Rocky sites, common. Endemic!

Flowering time:
Sommer-autumn

243

244

245

129

ASTERACEAE

246

Carlina involucrata

English:
Clustered Carline-thistle

Maltese:
Sajtun

Habitat/Distribution:
Garigue, steppes, common

Flowering time:
Summer

246

247

Cynara cardunculus

English:
Wild Artichoke

Maltese:
Qapoċċ tax-Xewk

Habitat/Distribution:
Garigue, steppes, common

Flowering time:
Spring-summer

Comments:
Height 20-100 cm

247

ASTERACEAE

248

Carthamus lanatus ssp. baeticus

English:
Woolly Safflower

Maltese:
Xewk ta' Kristu

Habitat/Distribution:
Steppes, disturbed sites, common

Flowering time:
Summer

ASTERACEAE

ASTERACEAE

249, 250

Palaeocyanus crassifolius (Centaurea crassifolia)

English:
Maltese Rock-centaury

Maltese:
Widnet il-Bahar

Habitat/Distribution:
Cliffs, valleys, rare. Endemic!

Flowering time:
Spring-summer

Comments:
Malta's national plant, cultivated as ornament

250

References

Aquilina J., 2000: English-Maltese Dictionary, four volumes. Midsea Books Ltd, Malta.

Borg, J., 1927: Descriptive Flora of the Maltese Islands. Government Printing Office, Malta. Authorized reprint 1976, O. Koeltz Science publisher Koenigstein, Germany.

Cassar, L.F. and D.T. Stevens, 2002: Coastal sand dunes under siege, Foundation for International Studies, Malta.

Haslam, S.M., P.D. Sell and P.A. Wolseley, 1977: A Flora of the Maltese Islands. Malta University Press, Malta.

Sultana J. (ed.), 1995: Flora u Fawna ta' Malta. Dipartiment għall-Ħarsien ta' l-Ambjent, Malta.

Sultana, J. and V. Falzon, 2002: Wildlife of the Maltese Islands. BirdLife Malta, Nature Trust, Malta.

Tutin et al., 1964-1980: Flora Europaea. Cambridge University Press, Great Britain. Haslam, S.M. and J. Borg, 1998: The river valleys of the Maltese Islands, Formatek Ltd, Malta.

Index of Scientific Names

A
Acanthaceae, 97
Acanthus mollis, 97
Aegilops geniculata, 45
Aegilops ovata, 45
Agavaceae, 41
Agave americana, 41
Aizoaceae, 60
Ajuga iva, 92
Alismataceae, 22
Alliaceae, 33
Allium cepa, 34
Allium commutatum, 34
Allium lojaconoi, 33
Allium porrum, 34
Allium subhirsutum, 34
Amaryllidaceae, 35
Anacamptis pyramidalis, 30
Anacamptis urvilleana, 31
Anacardiaceae, 85
Anthemis urvilleana, 117
Anthyllis hermanniae, 70
Anthyllis vulneraria ssp. maura, 71
Antirrhinum majus ssp. tortuosum, 98
Antirrhinum siculum, 98
Antirrhinum tortuosum, 98
Apiaceae, 110
Apium nodiflorum, 112
Araceae, 23
Arisarum vulgare, 23
Aristolochia longa, 21
Aristolochiaceae, 21
Arthrocnemum glaucum, 54
Arthrocnemum macrostachyum, 54
Arum italicum, 24
Arundo donax, 44
Asclepiadaceae, 89
Asparagaceae, 40
Asparagus aphyllus, 40
Asperula cynanchica, 88
Asphodelaceae, 28
Asphodelus aestivus, 28
Asphodelus fistulosus, 29
Aster squamatus, 121
Asteraceae, 116
Asteriscus aquaticus, 116
Asteriscus spinosus, 123

Atractylis gummifera, 128
Atriplex halimus, 56
Avena sterilis, 45

B
Bellis annua, 124
Bellis perennis, 124
Bellis sylvestris, 124
Blackstonia perfoliata, 90
Boraginaceae, 105
Borago officinalis, 107
Brassicaceae, 80

C
Cactaceae, 57
Caesalpiniaceae, 69
Cakile maritima, 82
Calendula arvensis, 122
Calendula suffruticosa ssp. fulgida, 122
Capparidaceae, 80
Capparis orientalis, 80
Capparis rupestris, 80
Capparis spinosa var. inermis, 80
Caprifoliaceae, 115
Carlina involucrata, 130
Carthamus lanatus ssp. baeticus, 131
Caryophyllaceae, 53
Centaurea crassifolia, 133
Centaurea melitensis, 129
Centaurea nicaeensis, 129
Centaurium erythraea, 90
Ceratonia siliqua, 69
Cerinthe major, 108
Chenopodiaceae, 54
Chiliadenus bocconei, 129
Chrysanthemum coronarium, 125
Cichorium intybus, 126
Cichorium spinosum, 126
Cistaceae, 84
Cistus creticus, 84
Cistus incanus ssp. creticus, 84
Clematis cirrhosa, 48
Clusiaceae, 63
Colchicaceae, 25
Colchicum cupani, 25
Convolvulus althaeoides, 102
Convolvulus elegantissimus, 102

Convolvulus oleifolius, 103
Convolvulus pentapetaloides, 103
Convolvulaceae, 102
Coridothymus capitatus, 94
Coronilla scorpioides, 72
Coronilla valentina var. glauca, 71
Crassulaceae, 62
Crataegus monogyna x azarolus, 77
Cremnophyton lanfrancoi, 55
Crithmum maritimum, 111
Crocósmia x aurea, 26
Crucianella rupestris, 88
Cucurbitaceae, 76
Cupressaceae, 18
Cupressus sempervirens var. horizontalis, 18
Cupularia viscosa, 117
Cuscuta epithymum, 104
Cuscutaceae, 104
Cynara cardunculus, 130
Cynodon dactylon, 45
Cynomoriaceae, 68
Cynomorium coccineum, 68
Cyperaceae, 42
Cyperus capitatus, 42

D
Damasonium alisma ssp. burgaei, 22
Damasonium bourgaei, 22
Darniella melitensis, 56
Daucus carota, 111
Dioscoreaceae, 37
Diplotaxis erucoides, 80
Diplotaxis tenuifolia, 81
Dipsacaceae, 114
Dittrichia viscosa, 117

E
Ecballium elaterium, 76
Echium italicum, 105
Echium parviflorum, 108
Echium plantagineum, 108
Erica multiflora, 86
Ericaceae, 86
Erodium malacoides, 60
Erodium moschatum, 60
Eryngium maritimum, 114

Euphorbia dendroides, 67
Euphorbia melitensis, 66
Euphorbia paralias, 65
Euphorbia pinea, 66
Euphorbiaceae, 65
Evax pygmaea, 119

F
Fabaceae, 70
Fagaceae, 79
Fagonia cretica, 61
Fedia cornucopiae, 115
Ferula communis, 113
Foeniculum vulgare, 113
Frankenia hirsuta, 52
Frankenia laevis var. hirsuta, 52
Frankeniaceae, 52
Freesia refracta, 25
Fumana thymifolia, 84

G
Galactites tomentosa, 127
Gentianaceae, 90
Geraniaceae, 60
Gladiolus italicus, 26
Gladiolus segetum, 26
Glaucium flavum, 51
Gynandriris sisyrinchium, 27

H
Hedysarum coronarium, 75
Helichrysum melitense, 119
Helichrysum rupestre var. melitense, 119
Holoschoenus vulgaris, 42
Hordeum leporinum, 43
Hordeum murinum, 43
Hyacinthaceae, 38
Hyoscyamus albus, 109
Hyoseris lucida, 118
Hyoseris radiata, 118
Hyoseris scabra, 118
Hyparrhenia hirta, 48
Hypericum aegyptiacum, 63
Hypericum crispum, 64
Hypericum pubescens, 63
Hypericum triquetrifolium, 64

I
Inula crithmoides, 116
Inula viscosa, 117
Iridaceae, 25
Iris sisyrinchium, 27

J
Juncaceae, 42
Juncus bufonius, 43
Juncus sorrentini, 42

K
Kickxia commutata, 99

L
Lagurus ovatus, 47
Lamiaceae, 91
Lathyrus clymenum, 75
Lauraceae, 20

Laurus nobilis, 20
Lavatera arborea, 85
Liliopsida, 21
Limonium virgatum, 59
Linaceae, 64
Linum strictum, 64
Linum trigynum, 64
Lobularia maritima, 82
Lonicera implexa, 115
Lotus cytisoides, 74
Lotus edulis, 75
Lotus ornithopodioides, 74
Lotus tetragonolobus, 74
Lygeum spartum, 44
Lythraceae, 79
Lythrum junceum, 79

M
Magnoliopsida, 20
Malva sylvestris, 84
Malvaceae, 84
Matthiola tricuspidata, 82
Matthiola incana ssp. melitensis, 81
Medicago polymorpha, 76
Mentha pulegium, 96
Mercurialis annua, 65
Mesembryanthemum nodiflorum, 60
Micromeria microphylla, 95
Mimosaceae, 68
Mirabilis jalapa, 57
Muscari comosum, 38
Muscari parviflorum, 38

N
Narcissus tazetta, 35
Nauplius aquaticus, 116
Nicotiana glauca, 109
Narcissus serotinus, 35
Nyctaginaceae, 57

O
Ononis natrix ssp. ramosissima, 73
Ophrys bertolonii, 32
Ophrys bombyliflora, 33
Ophrys fusca, 31
Ophrys speculum, 32
Ophrys vernixia, 32
Opuntia ficus-indica, 57
Orchidaceae, 29
Orchis collina, 29
Orchis conica, 30
Orchis coriophora ssp. fragrans, 30
Orchis lactea, 30
Orchis saccata, 29
Orobanchaceae, 100
Orobanche muteli, 100
Orobanche ramosa f. melitensis, 100
Oxalidaceae, 67
Oxalis pes-caprae, 67

P
Palaeocyanus crassifolius, 133
Pallenis spinosa, 123
Pancratium maritimum, 36
Papaveraceae, 51
Periploca angustifolia, 89

Phagnalon graecum ssp. ginzbergeri, 120
Phlomis fruticosa, 93
Picris echioides, 122
Pinaceae, 17
Pinus halepensis, 17
Pistacia lentiscus, 85
Plantaginaceae, 100
Plantago afra, 101
Plantago coronopus, 101
Plantago lagopus, 101
Plantago serraria, 100
Plumbaginaceae, 59
Poaceae, 43
Polygonaceae, 58
Polypogon monspeliensis, 46
Potentilla reptans, 77
Prasium majus, 94
Psoralea bituminosa, 73
Putoria calabrica, 87

Q
Quercus ilex, 79

R
Ranunculaceae, 48
Ranunculus bullatus, 49
Ranunculus ficaria, 49
Ranunculus muricatus, 49
Ranunculus saniculaefolius, 50
Ranunculus trichophyllus, 50
Reseda alba, 83
Resedaceae, 83
Rhamnaceae, 78
Rhamnus alaternus, 78
Rhamnus oleoides, 78
Ricinus communis, 65
Romulea ramiflora, 27
Rosa sempervirens, 77
Rosaceae, 76
Rosopsida, 48
Rubia peregrina, 89
Rubiaceae, 87
Rubus ulmifolius, 76
Rumex bucephalophorus, 58
Ruta chalepensis, 86
Rutaceae, 86

S
Salsola kali, 54
Salsola soda, 55
Salvia verbenaca, 95
Satureja microphylla, 95
Scabiosa maritima, 114
Scandix pecten-veneris, 112
Scilla autumnalis, 39
Scilla peruviana, 39
Scilla sicula, 38
Scirpus holoschoenus, 42
Scolymus hispanicus, 121
Scorpiurus muricatus, 72
Scrophulariaceae, 97
Senecio bicolor, 118
Senecio leucanthemifolius, 124
Silene colorata, 53
Smilacaceae, 28

Smilax aspera, 28
Smyrnium olusatrum, 110
Solanaceae, 109
Sporolobus arenarius, 43
Sporolobus pungens, 43

T
Tamaricaceae, 52
Tamarix africana, 52
Tamus communis, 37
Tetraclinis articulata, 19
Tetragonolobus purpureus, 74

Teucrium flavum, 92
Teucrium fruticans, 91
Thymbra capitata, 94
Thymus capitatus, 94
Tordylium apulum, 112
Triadenia aegyptica, 63
Tribulus terrestris, 61
Trifolium stellatum, 72

U
Urginea maritima, 39
Urginea pancration, 39

V
Valantia muralis, 88
Valerianaceae, 115
Verbascum sinuatum, 97
Verbena officinalis, 96
Verbenaceae, 96
Veronica anagallis-aquatica, 99
Veronica cymbalaria, 99
Vitex agnus-castus, 96

Z
Zygophyllaceae, 61

Index of English Names

A
African Tamarisk, 52
Aleppo Pine, 17
Alexanders, 110
American Agave, 41
Angiosperms, 7
Animated Oat, 45
Annual Beard-grass, 46
Annual Daisy, 124
Annual Hyoseris, 118
Annual Mercury, 65
Autumn Buttercup, 49
Autumn Grape Hyacinth, 38
Autumn Narcissus, 35
Autumn Squill, 39
Azarole, 77

B
Barbary Nut Iris, 27
Bay Laurel, 20
Beard-grass, 48
Bear's Breeches, 97
Bellardia, 98
Bermuda Grass, 45
Bertoloni's Orchid, 32
Black Bryony, 37
Blue Bindweed, 103
Blue Stonecrop, 62
Blue Water Speedwell, 99
Blue-leaved Acacia, 68
Boar Thistle, 127
Borage, 107
Bramble, 76
Branched Asphodel, 28, 29
Bristly Oxtongue, 122
Brown Orchid, 31
Buck's-horn Plantain, 101
Bumble Bee Orchid, 33
Bushy Restharrow, 73

C
Cape Sorrel, 67, 100
Caper, 80
Carob Tree, 69
Castor Oil Tree, 65
Creeping Cinquefoil, 77
Chaste Tree, 96

Chicory, 126
Clustered Carline-thistle, 130
Coast Ragwort, 124
Comb of Venus, 112
Common Birdsfoot Trefoil, 74
Common Centaury, 90
Common Freesea, 25
Common Kidney Vetch, 71
Common Mallow, 84
Common Pyramidal Orchid, 30
Common Smilax, 28
Creeping Loosestrife, 79
Crimson Pea, 75
Crisped St. John's-wort, 64
Crown Anemone, 48
Crown Daisy, 125
Cypress, 18

D
Dodder, 104
Dwarf Broomrape, 100

E
Eastern Phagnalon, 120
Edible Birdsfoot Trefoil, 75
Egyptian St. John's-wort, 63
Esparto Grass, 44
Evergreen Honeysuckle, 115
Evergreen Oak, 79
Evergreen Rose, 77
Evergreen Traveller's Joy, 48

F
Fagonia, 61
Fan-lipped Orchid, 29
Fennel, 113
Field Gladiolus, 26
Field Marigold, 122
French Daffodil, 35
Friar's Cowl, 23
Fringed Rue, 86

G
General's root, 68
Glandular Plantain, 101
Glandular Storksbill, 60
Goat Grass, 45

Golden Samphire, 116
Great Reed, 44
Great Sage, 93
Greater Snapdragon, 98
Green-flowered Birthwort, 21
Grey Birdsfoot Trefoil, 74
Ground Thistle, 128

H
Hairy Garlic, 34
Hairy Valantia, 88
Hare's-foot Plantain, 101
Hare's-tail Berley, 43
Hare's-tail Grass, 47
Hawthorn, 77
Hoary Rock-rose, 84
Honeywort, 108
Horn-of-plenty, 115

I
Italian Lords-and-Ladies, 24

L
Leeks, 34
Lentisk, 85
Lesser Celandine, 49
Lesser Crystal-plant, 60
Lesser Grape Hyacinth, 38
Liliopsida, 21

M
Mallow Bindweed, 102
Malta Fungus, 68
Maltese Cliff-orache, 55
Maltese Cross, 61
Maltese Dwarf Garlic, 33
Maltese Everlasting, 119
Maltese Fleabane, 129
Maltese Pyramidal Orchid, 31
Maltese Rock-centaury, 133
Maltese Salt-tree, 56
Maltese Savory, 95
Maltese Sea Chamomile, 117
Maltese Spurge, 66, 104
Maltese Star Thistle, 129
Maltese Stocks, 81
Marvel-of-Peru, 57

141

Mediterranean Buckthorn, 78
Mediterranean Fluellen, 99
Mediterranean Hartwort, 112
Mediterranean Heath, 86
Mediterranean Meadow Saffron, 25
Mediterranean Starfruit, 22
Mediterranean Stocks, 82
Mediterranean Stonecrop, 62
Mediterranean Thyme, 94, 104
Milky Orchid, 30
Mirror Orchid, 32
Montbretia, 26
Musk Storksbill, 60

N
Narrow-leaved Aster, 121

O
Olive-leaved Bindweed, 103
Olive-leaved Buckthorn, 78
Olive-leaved Germander, 91

P
Pale Bugloss, 105
Pale Speedwell, 99
Pennyroyal, 96
Perennial Hyoseris, 118
Perennial Wall-rocket, 81
Pine Spurge, 66
Pink Asphodel, 29
Pitch Clover, 73
Plantain Bugloss, 108
Prickly Pear, 57
Prickly Saltwort, 54
Procumbent Marshwort, 112
Pubescent St. John's-wort, 63
Pygmy Cudweed, 119

R
Red Campion, 53
Red Dock, 58
Rock Crosswort, 88
Round-headed Club-rush, 42

S
Sand Dropseed, 43
Sand Galingale, 42
Sand Oyster Thistle, 121
Sand-crocus, 27
Sandarac, 19
Sanicle-leaved Water Crowfoot, 50
Scented Bug Orchid, 30
Scilly Buttercup, 49
Scorpion-tail Vetch, 72
Sea Daffodil, 36
Sea Holly, 114
Sea Rocket, 82
Sea Samphire, 111
Sea Spurge, 65
Sea Squill, 39, 104
Sea-heath, 52
Seaside Ox-eye Daisy, 116
Seaside Sea-lavender, 59
Shepherd's needle, 112
Shrub Tobacco, 109
Shrubby Crown Vetch, 71
Shrubby Glasswort, 54
Shrubby Kidney Vetch, 70
Shrubby Marigold, 122
Shrubby Orache, 56
Sicilian Snapdragon, 98
Sicilian Squill, 38
Silvery Ragwort, 118
Slender Bindweed, 102
Small Lords-and-Ladies, 23
Small-flowered Bugloss, 108
Smooth-leaved Saltwort, 55
Southern Bugle, 92
Southern Daisy, 124
Southern Flax, 64
Southern Scabious, 114
Southern Spiny Asparagus, 40
Southern Star Thistle, 129
Spiny Chicory, 126
Spiny Ox-eye Daisy, 123
Squinancywort, 88
Squirting Cucumber, 76
Star Clover, 72
Sticky Fleabane, 117
Stinking Madder, 87
Sulla, 75
Sweet Alison, 82

T
Tassel Hyacinth, 38
Three-leaved Water Crowfoot, 50
Thyme-leaved Sun-rose, 84
Toad Rush, 42
Toothed Medick, 76
Toothed Plantain, 100
Tree Mallow, 85
Tree Spurge, 67

U
Upright Yellow Flax, 64

V
Vervain, 96

W
Wavy-leaved Mullein, 97
White Hedge-nettle, 94
White Henbane, 109
White Mignonette, 83
White Mustard, 80
Wild Artichoke, 130
Wild Carrot, 111
Wild Clary, 95
Wild Leek, 34
Wild Madder, 89
Winged Pea, 74
Wolfbane, 89
Woolly Safflower, 131

Y
Yellow Crown Vetch, 72
Yellow Germander, 92
Yellow Horned-poppy, 51
Yellow-wort, 90

Index of Maltese Names

A
Alaternu, 78
Almeridja tal-Blat, 54
Alizzari, 87
Anżalor, 77

B
Bżar tal-Patrijiet, 96
Bajtar tax-Xewk, 57
Barrum Vjola, 48
Basal il-Ħnieżer, 38
Beżżul il-Baqra, 62
Bebuna, 124
Bebuna tal-Baħar, 117
Berwieq, 28, 29
Biżbula Komuni, 101
Biżbula tal-Baħar, 101
Biżbula tas-Snien, 100
Bjanka, 56
Bjanka ta' l-Irdum, 55
Bordi tal-Ramel, 42
Borgħom Komuni, 92
Brijonja Sewda, 37
Brimba, 45
Broxka ta' Għawdex, 73
Bużbież, 111, 113
Budebbus ta' l-Ingliża, 100
Buqexrem, 96
Burikba, 65
Busieq, 25
Buttuniera, 82

Ċ
Ċentawrja Kbira, 90
Ċentawrja Safra, 90
Ċfolloq, 49, 50
Ċikwejra, 126
Ċipress Kannella, 18
Ċistu Roża, 84
Ċistu Żgħir, 84

D
Damażonju, 22
Denb il-Fenek, 47
Denb il-Liebru Kbir, 46
Denb il-Ħaruf, 83

Deru, 85
Dubbiena, 31
Dubbiena Kaħla, 32
Dubbiena ta' Bertoloni, 32

E
Evaks, 119
Erba Franka, 52
Erika, 86

F
Fagonja, 61
Faqqus il-Ħmir, 76
Fejġel, 86
Ferla, 113
Fexfiex Sufi, 63
Fexfiex ta' l-Irdum, 63
Fexfiex tar-Raba', 64
Fiġġiela Ħamra, 74
Fidloqqom, 107
Fjurdulis Salvaġġ, 27
Frawla Salvaġġa, 77
Friżja, 25

Ġ
Ġarġir Abjad, 80
Ġarġir Isfar, 81
Ġiżi ta' Malta, 81
Ġiżi tal-Baħar, 82
Ġilbiena tas-Serp, 75
Ġjaċint tal-Ħarifa, 38

G
Garni, 24
Garni tal-Pipi, 23
Gazzija, 68
Girlanda tal-Wied, 77

Għ
Għajn il-Baqra tax-Xatt, 116
Għajn il-Baqra Xewwikija, 123
Għansar, 38, 39
Għatba, 61
Għerq il-Ġeneral, 68
Għerq Sinjur, 68

Għolliq, 76

H
Ħabb il-Qamħ tar-Raba', 26
Ħalfa, 44
Ħannewija, 97
Ħarfur Kbir, 45
Ħatba Sewda, 70
Ħaxixa Ingliża, 67
Ħaxixa ta' l-Irmied, 55
Ħaxixa ta' l-Irmied Xewwikija, 54
Ħaxixa tal-Misk, 60
Ħaxixet it-Trieraħ, 112
Ħobbejża Komuni, 84
Ħobbejża tas-Siġra, 85
Hommejr, 57

I
Ingliża Sewda, 67

K
Kaħwiela, 48
Kappara, 80
Karfus il-Ħmir, 110
Karfus tal-ilma, 112
Kiesha, 48
Kittien Irqiq, 64
Kittien Isfar, 64
Koronilla, 71
Kristallina Komuni, 60
Kromb il-Baħar, 82
Kromb il-Baħar Isfar, 118
Kruċanella, 88
Kurrat Salvaġġ, 34

L
Leblieb tal-Blat, 103
Leblieb tal-Werqa Tleqq, 103
Leblieb tax-Xagħri, 102
Lellux, 125
Litrum ta' l-Ilma, 79
Lixka Komuni, 120
Lsien il-Fart, 108
Lsien il-Fart Abjad, 105
Lsien il-Fart Żgħir, 108
Lsien l-Għasfur, 53

M
Mammażejża, 109
Margerita Salvaġġa, 124
Maxxita, 112
Monbrezja, 26
Morra, 72
Moxt, 60

N
Naħla, 33
Narċis, 35
Narċis il-Baħar, 36
Narċis Imwaħħar, 35
Nefel Komuni, 76
Niġem, 43, 45
Nixxief Komuni, 43

O
Orkida Ħamra, 29
Orkida Piramidali, 30, 31
Orkida tat-Tikek, 30
Orkida Tfuħ, 30

P
Pankrazju, 36
Papoċċi Bajda, 98
Papoċċi Ħamra, 98
Pepprin Isfar, 51
Perlina Bajda, 98
Pittma, 104
Plejju, 96
Prinjoli, 17

Q
Qanfuda, 126
Qapoċċ tax-Xewk, 130
Qarn il-Mogħża, 115
Qarsajja, 58
Qasba Kbira, 44
Qniepen, 108
Qrempuċ, 74, 75

R
Ranġis, 35
Robbja, 87
Robbja Salvaġġa, 89

S
Sabbara ta' l-Amerika, 41
Sagħtar, 94
Sajtun, 130
Salsa Pajżana, 28
Salvja Salvaġġa, 95
Salvja tal-Madonna, 93
Sedum, 62
Sempreviva ta' Għawdex, 119
Settembrina Salvaġġa, 121
Siġra ta' l-Għargħar, 19
Siġra taż-Żnuber, 17
Siġra tal-Ballut, 79
Siġra tal-Bruk, 52
Siġra tal-Ħarir, 89
Siġra tal-Ħarrub, 69
Siġra tal-Virgi, 96
Siġra tar-Rand, 20
Siġra tar-Riċnu, 65
Sieq il-Ħamiema, 115
Silla, 75
Silla tal-Blat, 71
Silla tal-Mogħoż, 73
Simar tal-Boċċi, 42
Simar Żgħir, 42
Skabjoża, 114
Spraġ Xewwieki, 40
Suffejra Kbira, 122
Suffejra tar-Raba', 122

T
Tabakk tas-Swar, 109
Te Sqalli, 94
Tengħud Komuni, 66
Tengħud tar-Ramel, 65
Tengħud tas-Siġra, 67
Tengħud tax-Xagħri, 66
Tewm Irqiq ta' Malta, 33
Tewm Muswaf, 34
Tulliera Komuni, 117
Tulliera ta' Malta, 129

V
Veronika, 99
Veronika ta' l-Ilma, 99

W
Widna, 72
Widnet il-Baħar, 133

X
Xantkura, 92
Xatbet l-Andar, 97
Xatbet l-Art Vjola, 99
Xeħt l-Imħabba, 72
Xebb, 56
Xewk Abjad, 127
Xewk Isfar tar-Ramel, 121
Xewk Malti, 129
Xewk ta' Kristu, 131
Xewk ta' l-Għotba, 129
Xewk tal-Mixta, 128
Xewk tar-Ramel, 114
Xnien ta' l-Istilla, 72
Xorbett, 116
Xpakkapietra, 95

Ż
Żagħfran tal-Blat, 27
Żagħrun, 77
Żebbuġija, 91
Żerrigħ il-Brigħet, 101
Żigland, 118
Żiju, 78

Z
Zunnarija Salvaġġa, 111